DRUGS AND ALCOHOL POLICIES

Tricia Jackson BA, MSc (Personnel Management), MInstAm, FIPD is a freelance training and personnel consultant, specialising in employment law. Tricia has many years' experience as a generalist practitioner in both the public and private sectors. She is currently involved in tutoring on open learning and college-based IPD programmes, competence assessment, identifying and providing training solutions, personnel consultancy and representing clients at employment tribunals. Tricia is the co-author of the IPD's recommended textbook for the Certificate in Personnel Practice (*Personnel Practice*, M. Martin and T. Jackson, IPD, 1997) and author of *Smoking Policies*, IPD, 1999. Tricia lives in Weybridge, Surrey.

Other titles in the series

Bullying and Sexual Harassment
Tina Stephens

Creating a Staff Handbook
Clare Hogg

Induction
Alan Fowler

Part-Time Workers
Anna Allan and Lucy Daniels

Smoking Policies
Tricia Jackson

The Institute of Personnel and Development is the leading publisher of books and reports for personnel and training professionals, students, and all those concerned with the effective management and development of people at work. For details of all our titles, please contact the Publishing Department:

tel 020-8263 3387
fax 020-8263 3850
e-mail publish@ipd.co.uk

The catalogue of all IPD titles can be viewed on the IPD website:
www.ipd.co.uk

DRUGS AND ALCOHOL POLICIES

TRICIA JACKSON

INSTITUTE OF PERSONNEL AND DEVELOPMENT

First published in 1999

Design and typesetting by
Wyvern 21, Bristol

Printed in Great Britain by the Short Run Press, Exeter

British Library Cataloguing-in-Publication Data
A catalogue record for this book is available
from the British Library

ISBN 0-85292-811-4

INSTITUTE OF PERSONNEL
AND DEVELOPMENT

IPD House, Camp Road, Wimbledon, London SW19 4UX
Tel.: 020-8971 9000 Fax: 020-8263 3333
Registered office as above. Registered Charity No. 1038333
A company limited by guarantee. Registered in England No. 2931892

Contents

Acknowledgements

Many people and organisations have contributed to this book, either by way of formal contributions or through informal discussion. Special thanks are due to Richard Goff of IPD Publishing for his editorial input and words of encouragement. I would also like to express my grateful thanks to the following people who willingly gave their time to answer my questions. They are Sister Joanne Flooks and Nick Hawley of Denso Marston Ltd, Paul Radley of Rail-track and Jim Dunn and Tudor Price of British Steel plc.

What is a drugs policy?

- ✔ Introduction
- ✔ Benefits
- ✔ Reference

Introduction

Drug misuse is a growing problem in society and an increased number of employers are nowadays having to deal with its consequences. Those who have been proactive in developing policies are the best prepared. A policy on drugs can help an organisation to deal with drugs-related problems in the workplace in a considered way. The aim is similar to that of alcohol policies, ie significantly to reduce drug misuse in the workplace by heightening employee awareness of it.

There is a large range of drugs – some are illegal and others are not and their misuse can have varying effects. A simple categorisation is as follows:

- depressants – drugs that depress the nervous system, such as opiates, cannabis and barbiturates

- stimulants – drugs that stimulate the nervous system, such as cocaine, crack and amphetamines
- hallucinogens – drugs that alter mood and perception, such as LSD, magic mushrooms and designer drugs like ecstasy (which is also a stimulant)
- solvents – substances that alter mood and perception, including aerosol sprays, glues and butane gas. This category of drugs is not illegal but misuse can lead to health and efficiency problems.

Here the term 'drug misuse' is used to refer to employees who take illegal drugs or misuse prescribed drugs or substances such as solvents. You should note that, although drug misuse is not as widespread as alcohol misuse, it is a growing problem and many drug users are in employment. Furthermore, we are not just dealing with those employees who are dependent on drugs but also with many more who indulge in light or occasional use of drugs, ie recreational users.

In any event, your policy should differentiate between employees who are under the influence of drugs at work and those who indulge in criminal behaviour inside or outside the workplace (see Chapter 5 for further details).

Many of the problems associated with the misuse of drugs are similar to alcohol misuse and may be dealt with in the same way. For example drug dependence, like alcohol dependence, should be viewed as an illness and dealt with under the company's ill health or capability procedure. However, drugs differ from alcohol in that their use is not generally socially acceptable and may be illegal. This difference is also reflected in the tendency of employment tribunals to treat drugs-related offences less sympathetically than those involving alcohol. However, research findings are starting to indicate that there may be an underlying medical reason to explain drug taking. Thus employers would be advised to treat drug dependency in the same sympathetic manner as alcohol dependency, ie to provide assurances that employees who are identified as having misuse problems will be offered advice, treatment and confidentiality.

Oonagh Ryden, the IPD policy adviser on pay and employment conditions says that[1]:

> Companies should aim to prevent substance abuse by raising employees' awareness through education programmes on the health and safety risks. They should concentrate on encouraging people with a problem to seek help at an early stage before accidents happen or performance suffers. The emphasis should be on rehabilitation, with dismissal as a last resort.

Benefits

Drug misuse can affect the organisation in much the same way as alcohol misuse, ie it can lead to increased sickness and absenteeism, an increased risk of accidents, a deterioration in the quantity and quality of work and difficulties with working relationships. Furthermore, there may be an increased risk of theft or fraud as drug-dependent employees may become involved in criminal activity inside (or outside) the workplace in seeking to support their habit.

The benefits of implementing a drugs policy are therefore obvious and are essentially the same as those listed for alcohol policies at the end of Chapter 2.

Reference

1 IPD National Conference News Release. 16.11.98.

What is an alcohol policy?

- ✔ Introduction
- ✔ Benefits
- ✔ Reference

Introduction

Though the consumption of alcohol is an accepted part of social life in the UK, it can have significant implications for certain individuals and, where it affects employees' work performance, for employers. Alcohol policies vary considerably in scope and content but essentially they help employers to deal with alcohol-related problems in the workplace. The aim of an alcohol policy should be significantly to reduce alcohol misuse by increasing employees' awareness of its dangers.

We will be using the term 'alcohol misuse' to cover:

- employees who drink excessively on occasions, eg at the office party, but who are not physically or psychologically dependent on alcohol

- employees who consistently misuse alcohol, ie they are alcohol-dependent.

Alcohol policies usually differentiate between these problems. Employees in the first category tend to be dealt with under the disciplinary procedure but the health, absence and performance problems associated with the second category of employee indicate that an ill health or capability procedure is more appropriate. Good-practice employers are aware that this route is more likely to lead to the retention of (previously good) employees, especially if the problem is identified in its early stages so that the chances of recovery are greater.

Alcohol policies should be designed to encourage problem drinkers to acknowledge that they have a problem and to seek help. Employers should provide assurances that problem drinkers will be treated fairly and confidentially. The emphasis should be on seeking to identify staff who need help to tackle their addiction.

Alcohol misuse does not, however, just concern people with an alcohol dependency or those who occasionally overindulge. It is also to do with the culture of the organisation and the general attitudes of managers and employees to drinking practices. For instance, in your organisation is it commonplace for employees to drink alcohol at lunchtime? Is it expected that commercial and sales staff will drink alcohol when entertaining clients? Do employees openly swap stories about their hangovers on the morning after a social event? If the answer to these questions is yes, then not only is the performance and efficiency of your organisation likely to be adversely affected but risks to the health and safety of your employees and third parties are being taken.

Thus for an alcohol policy to be successfully introduced, some cultural changes may also be necessary. Without these, the health education aims of the policy may not be achieved.

Benefits

In a recent survey of 123 personnel directors, 90 per cent considered that alcohol consumption was a problem for their organisation (with 17 per cent describing it as a major problem)[1]. In particular they were concerned about:

- loss of productivity
- lateness and absenteeism
- safety issues
- the effect on team morale and employee relations
- bad behaviour or poor discipline
- adverse effects on the organisation's image and customer relations.

The figures provided in Chapter 3 highlight the scale of the problem in the UK and the associated costs to industry. No employer can expect to be spared these effects as it is statistically likely that even relatively small organisations will employ at least one individual with a drink problem.

Further, even when employees suffering from alcohol misuse do manage to attend work then 'sickness presence' can result. In such cases, employees are physically present but their performance is ineffectual and characterised by poor judgement, bad decision-making and an increased risk of accidents occurring. This is because, contrary to popular belief, alcohol is a depressant which slows down mental processes, making individuals less alert while giving them a false sense of confidence. In the event that such employees deal with outside parties in undertaking their duties, then damage may be done to the company's image and loss of potential business may also result.

There are many benefits to implementing an alcohol policy in the workplace and these will be explored in detail throughout this book. The arguments include:

- the employer's legal duties to ensure the health, safety and welfare of employees and third parties
- improvements in productivity, profitability, competitiveness, customer service, employee relations and corporate image
- reduced safety risks and absence levels
- the cost savings of supporting and retaining experienced staff who have a drink problem compared with dismissing them and recruiting and training new staff
- an increased likelihood that if a dismissal is necessary, it will be fair
- a better informed workforce, which in itself is an effective preventative measure
- the advantages to the organisation of a healthy and efficient workforce.

Reference

1 MORI research study conducted for the Health Education Authority (HEA). *Attitudes towards Alcohol in the Workplace.* February–March 1994.

Why do we need drug and alcohol policies?

- ✔ Combined or separate policies?
- ✔ The need for drug and alcohol policies
- ✔ Health and safety implications
- ✔ Ethical considerations
- ✔ Statistics on drugs and alcohol
  Drugs and alcohol – Drugs and alcohol policies – Employee assistance – Dismissals – Screening and testing
- ✔ The message
- ✔ References

Combined or separate policies?

Many organisations combine their drug and alcohol policies and produce a joint document. For the sake of simplicity we will assume that readers are designing and implementing combined policies throughout this book, but your organisation must decide whether a combined policy is appropriate or whether separate policies would be preferable. The choice depends on whether the aims of the drugs policy and the alcohol policy can be achieved. The conflicting arguments are set out below:

- Those in favour of a combined policy point to the fact that many of the issues involved are common to both drug and alcohol misuse and that the aims of the policies are broadly similar.
- Others argue that a combined policy covering the use of alcohol (which is legal) and drugs (which may be illegal) will be seen to be targeted at addiction and therefore irrelevant to the majority of the workforce.

Increasing employee awareness could alleviate some of the problems raised in the latter argument but employers should understand that, whatever the format, policies to cover both alcohol and drug misuse are necessary for a number of reasons. These are set out below.

The need for drug and alcohol policies

A report published in 1998 showed that one-third of employers covered by a survey of 275 union representatives have introduced a complete ban on indoor smoking at work and nearly half the employers in the survey have introduced smoking policies[1]. A much lower number of employers have established policies on the misuse of alcohol and fewer still on drug misuse. This is despite the benefits mentioned in Chapters 1 and 2 – such as improved performance, efficiency and morale.

Health and safety implications

There are a number of health and safety implications when employees are under the influence of drugs or alcohol while at work. These, and their legal context, are explored in Chapters 4 and 5.

Ethical considerations

We will be considering the controversial issues of screening and testing for drugs and alcohol usage – and the associated ethical factors – in Chapter 6. Here we will be concentrating on the scenario in which an organisation has discovered that one of its employees is dependent on alcohol or drugs. The employer may be tempted to take appropriate steps, in accordance with the disciplinary procedure, to dismiss that employee. However, good practice and ethical considerations dictate that using the workplace to tackle the addiction would be more appropriate for a number of reasons, as stated in the Institute of Personnel Management (IPM) Factsheet on anti-addiction programmes[2]:

- For most people, work is the most highly structured part of daily life. The signs and symptoms of developing an alcohol problem are constant, predictable and identifiable at the workplace. A fall-off in work performance is easily recognised.
- The workplace provides a unique opportunity for support and assistance from co-workers, who often notice things going wrong long before managers do.
- The possible loss of a job by the problem drinker can be used as an important factor in motivating change.

- The chances of recovery are much lower for problem drinkers who are not in work.
- It is more cost-effective for employers to encourage treatment than to sustain the cost of continuing poor performance, premature retirement or accidents. Most studies show that anti-addiction programmes more than pay for themselves.

This list covers alcohol dependency but is equally valid for drug addiction and reinforces the importance of a sympathetic rather than a punitive approach being taken by the employer.

In the next section we will consider some revealing statistics regarding alcohol, drugs and workplace policies in order to ascertain whether organisations are aware of the seriousness of these problems and are acting in a proactive fashion. The statistics reveal that a lot more needs to be done as the attitude of many employers is one of complacency and a lack of awareness.

Statistics on drugs and alcohol

Recent research from the USA suggests that:

- Employees who use drugs are a third less productive, 3.5 times more likely to injure themselves or a colleague in a workplace accident and 2.5 times more likely to be absent from work for eight days or more.
- 40 per cent of industrial fatalities and almost 75 per cent of industrial injuries may be linked to the use of drugs or alcohol.

At a joint Trades Union Congress (TUC), Alcohol Concern and Institute for the Study of Drug Dependence (ISDD) conference held in October 1998, delegates were informed of the following UK statistics[3]:

- Alcohol misuse costs employers an estimated £2 billion a year and drug misuse an estimated £800 million a year.
- Three out of four people with alcohol problems and one in four people seeking help with drugs are estimated to be in employment.
- One in four workplace accidents involve workers who have been drinking.
- Up to 14 million working days are lost each year through alcohol misuse, ie about 3 to 5 per cent of all absence.

A recent Incomes Data Services (IDS) Study provides the following information[4]:

- Consumption of alcohol per head rose by over 70 per cent from 1960 to 1996.
- 29 per cent of full-time employees have used illegal drugs and 48 per cent of 16- to 24-year-olds have used drugs at some time.

The 1998 survey of 1,800 UK personnel professionals carried out by the Reward Group for the IPD provides the following statistics[5]:

Drugs and alcohol

- 46 per cent of firms had received reports of alcohol misuse by staff in the last year (compared to 35 per cent in 1996).

- 18 per cent of firms had received reports of illegal drug-taking by staff in the last year (compared to 15 per cent in 1996).

Of the firms that had received reports, the problems emanated from a variety and combination of different sources:

Reports of Drug and Alcohol Misuse

Sources	Alcohol %	Drugs %
Deteriorating job performance	83	64
Deteriorating working relationships with co-workers	69	57
Harmed relationships with clients	40	27
Workplace accidents	14	13

Drugs and alcohol policies

- 39 per cent of respondents said that their organisation did not have an alcohol policy and 47 per cent said that their organisation did not have a drugs policy.
- Only 17 per cent of respondents reported an alcohol awareness policy and only 15 per cent a drug awareness policy.
- Only 14 per cent and 13 per cent respectively of respondents said that their managers/supervisors are trained to recognise drink misuse and the signs of illegal drug-taking.

Employee assistance

- 92 per cent of firms encouraged employees with an alcohol problem to seek help and counselling and 51 per cent allowed time off for rehabilitation.
- 81 per cent of firms encouraged employees who used illegal drugs to seek help and counselling but only 38 per cent allowed time off for rehabilitation.

Dismissals

- 18 per cent of firms always dismissed employees who abused alcohol and 31 per cent always dismissed employees who used illegal drugs, regardless of the nature of the job.

Screening and testing

- 9 per cent of firms had a policy of pre-employment screening for alcohol misuse and the same per cent had a policy of pre-employment screening for illegal drug misuse.
- 4 per cent had a policy of random alcohol testing and 5 per cent had a policy of random drug testing during employment.

The message

The message is clear. Employers have a duty to provide a safe and healthy working environment and to take care of their employees. The IPD urges employers to develop workplace policies that address drug and alcohol misuse. The aim should be to raise awareness of these issues and to encourage employees with problems to seek help.

Complacency is not an option, as this perpetuates the status quo. Such an approach will have a detrimental effect on an organisation's competitiveness, quality of service provision and profitability – as well as threatening the health, careers and lives of individuals suffering from drug and alcohol dependency.

References

1 Bargaining Report 183. *Clearing the Air at Work*. May 1998. pp8–10.
2 IPM Factsheet 20. *Anti-Addiction Programmes*. August 1989. pp1–2. (No longer available.)
3 TUC Press Release. 12.10.98. *Drink and Drug Misuse Cost Employers £3 billion*. Drugs and Alcohol Conference.
4 IDS Study. *Alcohol and drugs policies*. No. 652. August 1998. p2.
5 IPD Reward Group Salary Survey. *Drugs and Alcohol in the Workplace*. 1998. pp1–31.

What are the signs, effects and causes of drug and alcohol misuse?

- ☑ Signs
- ☑ Effects
- ☑ Causes
- ☑ Cautionary note
- ☑ References

Signs

Drug and alcohol misuse pervades all sections and levels of society. There are variations in the proportion of users in different age ranges, but the problem cuts across educational, class and ethnic boundaries. With regard to excessive drinking, high-risk occupations are, according to the IPD, characterised by the following factors[1]:

- availability of alcohol at work (eg in the drinks trade)
- social pressure to drink at work (eg among servicemen and journalists)
- separation from normal social and sexual

relationships (eg seafarers and commercial travellers)
- freedom from supervision (eg professionals such as doctors and lawyers)
- very high and low income levels
- occupational stress caused by danger (eg in the armed services), by responsibility (eg nurses, doctors, lawyers, managers), by lack of job security and boredom caused by underoccupation.

As you can see, this list covers a wide range of industrial sectors. Similarly, drug use in Britain has changed markedly over recent years and employers should not have stereotypical images of drug addicts and pushers in mind when designing their drug policies. There are many recreational drug users who are employed in a wide range of managerial, office and shopfloor posts across all industrial sectors.

There are many signs of drug and alcohol misuse. A recent IDS Study lists them as shown in the table opposite[2]. Employers should also be aware of the implications of discovering items associated with drug-taking, eg needles, syringes, paper twists, small mirrors, scorched pieces of tin foil and homemade pipes.

When looking for signs of drug or alcohol misuse, employers should note that some of the symptoms are very similar to those of a range of medical conditions, notably diabetes and epilepsy. The correct response is to make a full investigation of all the circumstances, including gathering of medical evidence, before making any decisions. There may also be contributory factors such as mental illness, depression or work-related stress which would need to be taken into account before determining a course of action.

Possible signs of drug and alcohol misuse

Reduced work performance characterised by:
- confusion
- lack of judgement
- impaired memory
- difficulty in concentrating on work
- periods of high and low productivity.

Absenteeism and timekeeping:
- poor timekeeping
- increased absence
- peculiar and increasingly improbable excuses for lateness and absence.

Personality changes:
- sudden mood changes
- irritability and aggression
- overreaction to criticism
- friction with colleagues.

Physical signs:
- smelling of alcohol
- loss of appetite
- unkempt appearance
- lack of hygiene.

Feeding the addiction:
- attempting to borrow money from colleagues
- dishonesty.

In any event, an early intervention is preferable so that employees who suffer from drug or alcohol dependency can be helped and, hopefully, rehabilitated. This is even more crucial in the former case, as the use of some drugs can affect physical and mental health more rapidly than does alcohol abuse.

We will now consider the effects on health of drug and alcohol misuse.

Effects

We have already discussed the short-term effects of drinking alcohol – ie it impairs judgement, making the performance of complex mental and physical tasks more difficult. The longer term effects of excessive drinking include serious liver diseases such as hepatitis, cirrhosis and cancer. Other long-term ill-health effects include:

- raised blood pressure
- cancer of the mouth, throat, stomach and breast
- stomach disorders
- sexual problems
- obesity
- depression
- psychiatric disorders.

Further, combined with drugs such as barbiturates, tranquillisers, stimulants and solvents, the effects can be more potent and possibly life-threatening. People who smoke and drink are also at an increased risk of dying from cancer because alcohol enhances the action of cancer-producing agents, such as tobacco smoke.

The picture is just as gloomy when we consider drug misuse. The short-term effects of taking drugs are impaired co-ordination, longer reaction time and an inability to maintain attention, with obvious implications for the performance of work tasks.

In the longer term, drug users may become dependent physically or psychologically on a drug. Drugs may be adulterated or an individual could mistakenly take an overdose with unpredictable results, possibly fatal ones. Injecting carries a number of risks, notably infection from non-sterile needles (including hepatitis and HIV), abscesses, thrombosis, gangrene and blood poisoning. Finally, the combined effects of taking drugs together (or with alcohol) are that their effect lasts much longer, they are more likely to be dangerous and again may have fatal consequences.

Next we examine the causes of drug and alcohol misuse. These have already been touched upon in the first section of this chapter.

Causes

We will look at alcohol misuse first. There are a number of potential causes, which may originate in the workplace or the domestic situation. In a *Croner's Guide*, they are listed as[3]:

- the volume of work
- monotonous or boring work
- unsocial or irregular hours
- under or over-promotion
- workplace stress
- access to alcohol at work
- working alone or without supervision
- marital or family problems

- family illness or bereavement
- financial difficulties
- pressure from colleagues or friends.

The *Croner's Guide* provides a similar list covering the causes of drug taking[4]:

- pressure from friends and colleagues
- curiosity
- too much or too little work
- lack of management direction and supervision
- unsocial or irregular hours
- monotonous work
- stress, depression or anxiety.

Cautionary note

In Chapter 7 we stress the importance of training managers so that they are confident in implementing the company's drugs and alcohol policy. As well as enforcing the new rules when infringements occur, managers have an important role to play in identifying the signs of drug or alcohol misuse. Most managers will be reasonably comfortable in deciding whether employees have overindulged in their consumption of alcohol but the signs of drug misuse are much more difficult to detect. This fact needs to be recognised in the training provision and managers should be encouraged to seek medical advice so that the signs described above are not confused with the symptoms of other physical or mental conditions.

Another important stage in the process of developing a new policy is that of educating the workforce. One of the main messages that must be communicated is that turning a blind eye or colluding by covering up an individual's drugs or drink problem is not acceptable. As we saw above, the long term effects of drug and alcohol misuse are dangerous and can certainly threaten careers, if not wreck whole lives. A preventative approach, whereby problems are tackled in the early stages, should be strongly encouraged. In this way treatment of the individual's health problem has a better chance of success and the organisation may be able to retain an experienced employee.

References

1 IPM Factsheet 20. *Anti-Addiction Programmes*. August 1989. p1. (No Longer available.)
2 IDS Study 652. *Alcohol and Drugs Policies*. August 1998. p3.
3 BAKER N. L. *Croner's Guide to Handling Sensitive Issues in the Workplace*. Croner Publications Ltd. 1997. p68.
4 BAKER N. L. *Croner's Guide to Handling Sensitive Issues in the Workplace*. Croner Publications Ltd. 1997. p76.

5

What rights and duties does the law provide?

- The legal position
- Health and safety – common-law duties
- Health and safety – statutory provisions
- Drugs and alcohol – misconduct or medical problem?
- Drugs and alcohol – disciplinary and capability procedures
- Drugs, alcohol and unfair dismissal
- Drugs, alcohol and the contract of employment
- Drugs, alcohol and discrimination
- National initiatives
- Reference

The legal position

The legal position for employers regarding drugs and alcohol in the workplace is a complex one. They have to consider:

- their own statutory and common-law duties of care to their employees and to third parties
- the statutory and common-law responsibilities of their employees, ie employees who are under the influence of drugs or drink and employees who

fail to report colleagues who are under the influence of drugs or drink
- the limited but specific regulation of the consumption of drugs and drink in the workplace
- whether, in handling a drugs- or drink-related offence, they are dealing with a conduct or a capability issue, for which different procedures apply
- the legal implications of drug misuse in the workplace, ie it is a criminal offence to possess, give away or sell an illegal drug
- whether disciplinary action, including dismissal, is justified for activities that occurred outside the workplace
- the legal implications of screening job applicants and testing existing employees for drug and alcohol misuse. (This issue will be dealt with in Chapter 6.)
- the impact on the employee's contract of employment of the introduction of a new workplace policy on drugs and alcohol.

There are therefore a number of principal legal areas of concern that must be taken into account when introducing a drugs and alcohol policy and these are considered below.

Health and safety – common-law duties

Common law imposes a duty on employers to take reasonable care of the health and safety of their employees. This involves providing a safe system of work, effective supervision and competent fellow employees. Turning a blind eye to the signs of drug or alcohol misuse could be deemed to be a breach of this duty. If other employees or third parties suffer an injury as a result, employers will find that they are liable to pay damages.

Employees also have responsibilities in relation to health and safety. If they fail to carry out their work with reasonable care because they are under the influence of drugs or drink and damage or injury results, they could in theory be sued for negligence. In practice, it is more likely that the employer will be sued because employers are vicariously liable for the actions of their employees committed during the course of employment.

These duties of care are echoed in the statutory provisions.

Health and safety - statutory provisions

The Health and Safety at Work Act (HASAW) 1974 places a general duty on employers to ensure the health, safety and welfare of their employees. This includes providing and maintaining a safe place of work and a safe system of work together with adequate supervision. Employers must ensure that employees do not injure themselves and/or constitute a danger to others. In the context of drug and alcohol misuse by employees, the likelihood of this occurring increases when they use machinery, equipment or vehicles. Breach of this duty is a criminal offence.

Further, HASAW places employers under a statutory duty to conduct their undertakings in a way which ensures, so far as is reasonably practicable, that people who are affected by the operation of those undertakings are not exposed to health and safety risks. This concerns failure to protect third parties, which is also a criminal offence.

With regard to employees, HASAW imposes a statutory obligation on them to take reasonable care for their own health and safety and that of others who may be affected by their actions or omissions at work. They must also co-operate with their employer to enable the employer to comply with its own duties under the legislation. It is arguable that there is a legal obligation on employees to inform their employer if they know that a colleague is under the influence of drink or drugs. Further, employees lawfully taking drugs on medical advice may also be at risk or pose a danger to others. There is therefore a duty on those employees to warn the employer of that risk.

Are there any other relevant statutory provisions? In certain sectors of industry, particularly those where safety considerations are of paramount importance, eg the offshore industry, the employer may have strict rules concerning drug and alcohol misuse but only limited statutory regulation is in place. In the transport industry, however, there is specific legislation to control the misuse both of drugs and alcohol. The Transport and Works Act 1992 makes it a criminal offence for certain workers to be unfit through drink and/or drugs while working on railways, tramways or other guided systems. The employer would also be guilty of a criminal offence unless it could be shown that 'all due diligence' had been exercised in trying to prevent those offences being committed.

Further, three sets of regulations prohibit the consumption of alcohol by employees in specific situations. They are:

- the Work in Compressed Air Special Regulations 1958
- the Control of Lead at Work Regulations 1980
- the Ionising Radiations Regulations 1985.

There is little specific regulation of the use of drugs in the workplace but the Misuse of Drugs Act 1971 makes it an offence to possess, supply, offer to supply or produce controlled drugs without authorisation. The Act lists the drugs that are subject to control and classifies them according to the perceived danger.

Classified drugs	
Class A	including heroin, opium derivatives, cocaine, LSD and ecstasy
Class B	including barbiturates, cannabis and oral preparations of amphetamines
Class C	including tranquillisers such as valium, sleeping pills and less harmful amphetamines

It is also an offence for employers knowingly to allow controlled drugs to be supplied, produced, used or simply held on their premises. A criminal prosecution could result and if an offence has been deemed to be committed with the consent or connivance of or owing to the neglect of a director, manager, secretary or other similar officer of the organisation, that officer is also guilty of a criminal offence.

Drugs and alcohol - misconduct or medical problem?

The Advisory, Conciliation and Arbitration Service (ACAS) in its advisory handbook, *Discipline at Work*, suggests that employers treat drug and alcohol misuse as a medical problem rather than a disciplinary matter[1]. They point out that managers may be able to identify symptoms such as poor performance, mood changes, anxiety, depression and deterioration in personal hygiene as potential signs of misuse. Their concerns should be discussed confidentially with the employee. The next step would be to obtain medical advice (with the employee's consent) in order to ascertain whether there is an underlying health problem. Contributory factors such as severe stress and mental illness should also be explored.

There are occasions, however, when the disciplinary approach would still be appropriate, eg serious incidents of drunkenness at work or the storage of drugs on premises. Such incidents may be considered to be examples of gross misconduct (and should be listed as such in the disciplinary rules). Summary dismissal (dismissal without notice or payment in lieu of notice) may be the inevitable result.

Can we be specific about which approach is appropriate? We said in the introduction to Chapter 1 that alcohol policies tend to differentiate between employees who are dependent and those who are not. Thus, in simple terms, when incidents occur, the correct approach for non-dependent drinkers is to use the disciplinary procedure, whereas a capability procedure is more appropriate for dependent drinkers. The steps applicable to these two approaches are set out in the next section.

Does this hold true for drug misuse? Awareness of the causes and treatment of drug misuse has risen substantially in recent years, although traditionally tribunals have tended to regard drug-related dismissals as misconduct issues. Current views on good practice dictate that employers should be willing to differentiate between recreational users where conduct is the issue, and drug dependency where the employee's health is the issue.

In reality the choice may not be this simple, as it is not always obvious whether you are dealing with dependency or occasional overindulgence. Another important factor that applies to drink and drug misuse is whether the employee in question works in a safety-critical post or industry, in which case disciplinary action may be the only option.

We will now refer to a fictional organisation in order to provide further clarification. We will assume that our fictional organisation's policy states that:

- drug and alcohol dependency will be treated as a medical problem
- breaches of the policy which are not related to dependent behaviour will be treated as misconduct
- stricter penalties are reserved for non-compliance with the rules for employees in safety-critical posts
- an employee assistance programme is provided to help employees who are alcohol or drug dependent.

The flowchart on page 32 sets out the management referral process applicable here.

Correct management of drug- or alcohol-related incidents

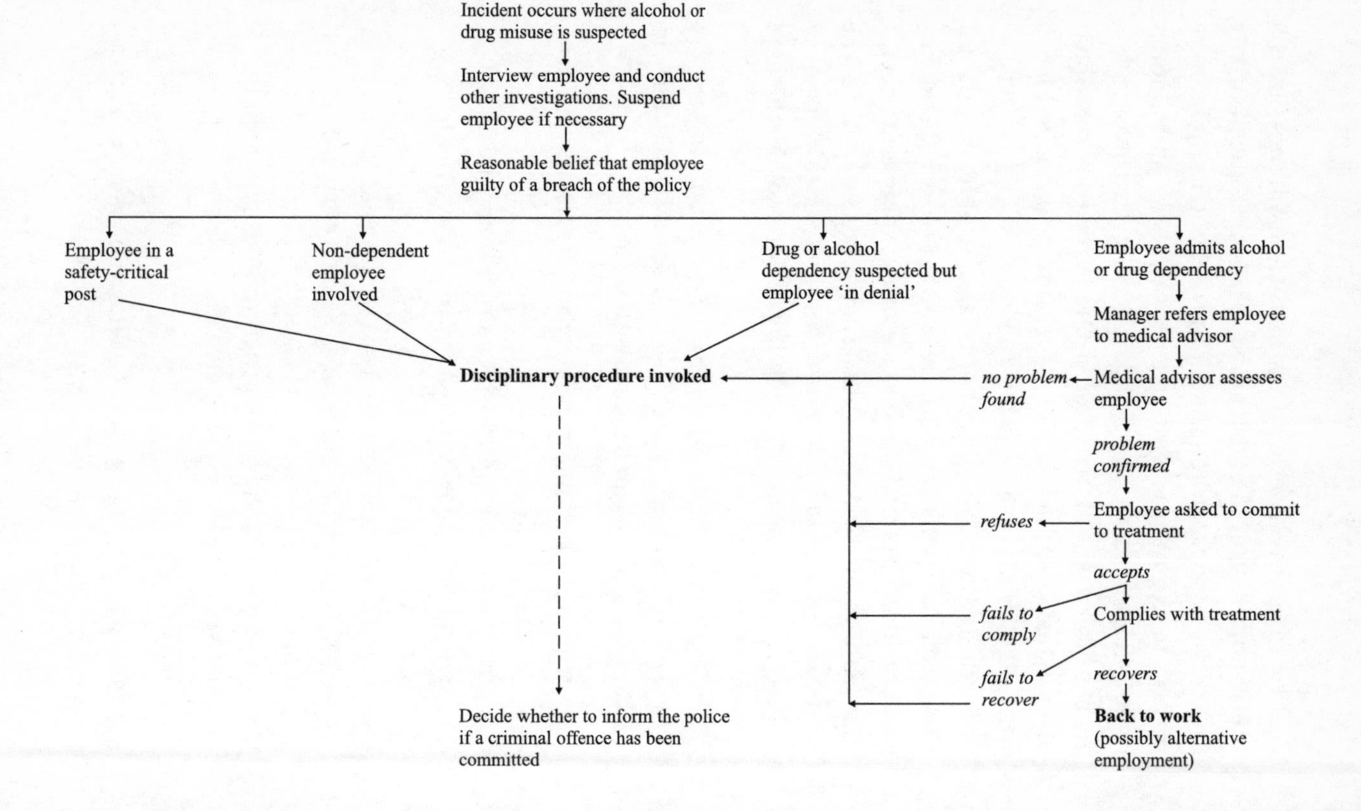

Note: where a criminal offence has been committed (usually) outside of the workplace, and a conviction is pending, eg for possession of illegal drugs or a drink-driving offence, this does not necessarily mean that the employer should dismiss the employee concerned. Each case should be considered on its merits and the key is to decide whether the employee's conduct warrants disciplinary action because of the employment implications. (See the unfair dismissal section below.)

Drugs and alcohol - disciplinary and capability procedures

Disciplinary procedures help to ensure that disciplinary rules are observed and standards are maintained. There are several key elements to a fair procedure including:

- the need for a proper investigation of the facts
- the opportunity for the employee to state his or her case
- the right of the employee to be accompanied by a colleague or representative
- the right of appeal against any disciplinary action.

Capability procedures follow a rather different route, although both may have the same conclusion – ie dismissal of the employee. The key elements of a capability procedure are:

- consultation with the employee
- medical investigation
- consideration, where appropriate, of alternative employment.

In tribunal proceedings, the panel will consider whether these three elements were provided by employers. We have already stressed the importance of a full medical investigation. With regard to consultation, this should include discussions with the employee as soon as the drug- or alcohol-related problem has been identified and throughout the period of treatment. This personal contact is essential so that the employee's opinion of his or her condition, chances of recovery and the suitability of alternative employment can be taken into consideration. Most importantly, the employee should be informed if the stage when dismissal may be an option, eg following a relapse, is approaching.

In summary, before an employer dismisses an employee for drug- or alcohol-related reasons, it should be clear whether the issue is one of misconduct (to be dealt with under the disciplinary procedure) or of ill health (to be dealt with under the capability procedure). This is because, in tribunal proceedings, the panel will consider whether the employer acted reasonably in treating its reason as a sufficient one for dismissing the employee, and whether it followed a reasonable procedure. However, should the employer set off on one path but then discover that it is the wrong one, it is not too late to alter its approach. An example of this is provided in the case on page 35.

Case-study	
Case-law	**Key learning-points**
Martin v *British Railways Board* 362/91 (EAT) An employee was dismissed for being drunk at work after his shift manager noticed that he was slurring his words and had glazed eyes. His internal British Railways appeal failed although, in his defence, the employee said he had been suffering from hypertension.	The tribunal and the EAT decided that the dismissal was unfair because, after the appeal hearing, the employer should have carried out further investigations to ascertain whether there was a health problem.

Drugs, alcohol and unfair dismissal

The Employment Rights Act (ERA) 1996 sets out the five potentially fair reasons for dismissal:

- conduct
- capability or qualifications
- redundancy
- the fact that the continued employment of the employee would have involved a contravention of some statutory duty
- 'some other substantial reason'.

As we saw above, we are dealing in the main with conduct and capability dismissals. It is worth noting, however, the situations listed overleaf:

- A drink-driving conviction might result in an employer deciding, where the post involves a substantial amount of driving, to dismiss on the grounds of a contravention of statutory duty. Transportation options and the possibility of alternative employment need to be fully discussed with the employee before reaching this conclusion.
- A charge or conviction for outside criminal behaviour may lead to dismissal on conduct grounds or 'for some other substantial reason', eg for attracting bad publicity to the organisation. In both cases, the employer would need to establish a reasonable belief, following investigation, that the employee was guilty of the misconduct in question. Further, the employee's off-duty behaviour would need to reflect on the employer/employee relationship (see the Focus DIY case on page 38).

Both avenues are fraught with danger for the employer, as are the main ones that we have been considering: conduct and capability reasons.

Following the implementation of a drugs and alcohol policy, there are a number of principles that employers need to adhere to if they want to ensure that dismissals are fair. These principles have been established by the ACAS code of practice on disciplinary practice and procedures in employment (revised in 1997) and case-law. Essentially the employer must ensure that:

- employees are aware of the policy
- the policy is a reasonable one that takes account of the practicalities of the workplace and the nature of the business

- the policy is applied in a considerate manner
- if a breach of the policy is, in the employer's view, potentially a dismissable offence then a breach will be treated as a serious disciplinary matter (NB it would also be wise to include a breach of the policy as an example of gross misconduct in the disciplinary rules/procedure – see Chapter 7)
- breaches are handled in a consistent manner
- assistance is given to employees who are drug or alcohol dependent
- due consideration is paid to the needs of the individual
- the use of the grievance procedure is encouraged
- any dismissals are procedurally fair
- dismissal is a reasonable response in all the circumstances.

Case-study	
Case-law	**Key learning-points**
John Walker & Sons v *Walker* 725/83 (EAT) An employee was dismissed for gross misconduct when he was found to be under the influence of drink at work. The company claimed that it had a rigid and well-known rule that anyone found to be drunk would be automatically dismissed.	The tribunal and the EAT found that there was no such rule in the company's disciplinary code and that, though the employee was under the influence of alcohol, he was not so drunk that he was unfit to carry out his duties. Furthermore his 20 years' service should have been taken into account. They concluded that, without a clear and notorious rule, dismissal was not a reasonable response.

Focus DIY v *Nicholson* 225/94 (EAT) A senior employee attended a company-organised evening function at a hotel. Along with some other staff, she smoked cannabis in front of other staff. She was dismissed on the basis that she had represented the company in public and had undermined her position at work by offending other staff.	The tribunal felt that the employer had not investigated the circumstances sufficiently and decided that the dismissal had been unfair. The tribunal observed that young people generally had a different attitude towards the use of cannabis nowadays. This decision was overturned by the EAT. They felt that the dismissal was fair and that the tribunal should not have substituted its own view for that of the employer.
Paul v *East Surrey District Health Authority* [1995] IRLR 305 (CA) A psychiatric nurse at a mental hospital drank whisky while on duty, became aggressive and began abusing other staff members and disturbing the patients. He had to be restrained and the police were called. He was dismissed for gross misconduct as he was deemed to be in dereliction of his duty.	The tribunal agreed that he had been in breach of the hospital's rule that no alcohol should be consumed on duty without permission. However, they found that other employees found to be drunk on duty had not been dismissed. Their finding of unfair dismissal was overturned at the EAT, and this was upheld at the Court of Appeal (CA), because the cases referred to were not comparable.
Williams and others v *Whitbread Beer Co* [1996] IRLR 560 (CA) Three employees were dismissed for drunken conduct following a company behavioural skills course run in one of the employer's hotels.	The CA upheld the original tribunal ruling that the employees had been unfairly dismissed on the basis that the misconduct took place outside work. Further, the employees all had long service and the employer had provided a free bar all evening.

Drugs, alcohol and the contract of employment

The contract of employment regulates the relationship between the employer and the employee. The terms of the contract have many sources, eg statute law, common law, collective agreements, staff handbooks and custom and practice. The terms of the contract will either be express ones or implied ones. The former are those terms which are written down or expressly agreed by the two parties to the contract and the latter are implied into the contract from a variety of sources including statute, custom and practice and common law.

The introduction of a policy on drugs and alcohol will impact on the contract of employment in two major ways, by necessitating a change to:

- disciplinary rules
- terms and conditions of employment.

In the former case it is usually possible to put these changes into effect without the consent of employees, but they must be notified of the change. In the latter case such changes can only be effected with the employees' consent, or by terminating their contracts on due notice and offering revised terms and conditions. In any event, employees should be informed of the need for change and fully consulted in order to avoid a finding of unfair dismissal at employment tribunals.

On a related note, claims of constructive dismissal stem from employees alleging that they were forced to resign as a result of some action on the part of the employer which constituted a breach of the contract of employment. How might employers breach the contract of employment in relation to drugs and alcohol in the workplace? They might be deemed to have breached the contract by implementing a new policy without consultation with employees, as discussed above, or by enforcing the terms of the policy, eg in relation to drugs and alcohol testing.

Employers should be aware that, in the absence of an express term, the courts and tribunals have been reluctant to imply into contracts of employment a right to allow employers to test or search employees for drugs or alcohol. If an employee were forced to submit to a search or test then this would amount to a breach of contract (as well as a criminal offence of assault). The employee might decide to resign and claim constructive dismissal (see Chapter 6 for further information).

Drugs, alcohol and discrimination

It should be noted that a refusal to employ individuals with a drugs or drink problem is not contrary to any of the discriminations acts and, in fact, the Disability Discrimination Act (DDA) 1995 expressly excludes:

> addiction to or dependency on alcohol, nicotine, or any other substance (other than as a result of the substance being medically prescribed).

Thus, with regard to pre-employment screening for alcohol and drugs, it would not be discriminatory to reject a potential candidate if he or she received a positive test result. Employers must take care, however, not to handle the application of screening and testing in a discriminatory way, ie to direct it at certain individuals/sectors of the population on the assumption that they will be at a greater risk of being alcohol or drug dependent. (See Chapter 6 for more information.)

However, existing or potential employees who engage in drug or alcohol misuse may suffer a deterioration in their health to the extent that the medical diagnosis suggests a long-term and substantial physical impairment. At this stage, it is possible that they could be classified as 'disabled' within the terms of the Act. This must be borne in mind so that decisions relating to the recruitment or continuing employment of the individual are taken in the light of medical evidence and the provisions of the DDA.

National initiatives

In response to the growth in the drug culture and concerns about increases in alcohol consumption in the UK, a number of government and pressure group initiatives have been launched in recent years:

- In 1995 the government White Paper *Tackling Drugs Together* emphasised the need for a multidisciplinary approach, involving the public and private sectors, to tackle the growing problem of illegal drugs in society.
- In 1998, the Labour government launched a 10-year strategy which develops the work of the white paper and encourages employers to initiate workplace policies.
- Also in 1998, a forum of employers, employees, trade union bodies and the Health and Safety Executive (HSE) mounted a campaign called 'Drugs and the Workplace' to help firms develop workplace policies.

- The ISDD and Alcohol Concern have been working with other interested parties to consider the effects of drink and drugs in the workplace and provide guidance on workplace policies.

Reference

1 ACAS Advisory Handbook. *Discipline at Work*. Revised February 1999. p44.

What about screening and testing?

- ✔ Prevalence in the UK
- ✔ Terminology
- ✔ Pre-employment screening
- ✔ Testing during employment
- ✔ Handling the results
- ✔ Shortcomings
- ✔ Ensure agreements

Prevalence in the UK

Legislative developments in the USA in recent years have led to widespread implementation of drug and alcohol screening programmes by employers. This practice is not as common in the UK but it is growing, particularly in organisations that come under the remit of the Transport and Works Act 1992. Some organisations in the UK have been testing employees for years and these include safety-critical industries such as oil production, shipping and the armed services. Testing has also been introduced into other sectors where impaired performance has high financial risks attached to it, for example among workers on the City of London money markets.

Testing for drugs and alcohol in the workplace has always been a controversial issue. Opponents claim that it is an infringement of civil liberties and human rights. Supporters stress the obvious health and safety risks attached to allowing individuals who are under the influence of drink or drugs carry out their normal day-to-day duties. Further they point to evidence from the USA that suggests that testing can reduce drug usage.

Terminology

The terms screening and testing tend to be used interchangeably but here we will, in the main, use 'screening' when the recipients are potential employees and 'testing' when referring to existing employees.

Accredited laboratory services test for drugs and alcohol use by taking either a blood, urine or hair sample from employees or potential employees. Many also offer an alcohol breathalyser test service.

Pre-employment screening

Generally speaking, employers are entitled to require potential employees to submit to tests. Breathalyser or other tests to detect alcohol for potential employees are less common than screening for drugs usage. Further they have a limited usefulness (see the section on shortcomings, pages 47 to 49). Medical questionnaires and examinations may, however, reveal whether individuals have a drinking problem.

We mentioned in Chapter 5 that the DDA specifically excludes addiction to drugs or alcohol from the definition of disability (though the long-term health effects may fall within the remit of this legislation). It would not therefore be discriminatory to reject for employment an individual who is dependent on drugs or alcohol. However, care needs to be taken when screening (and testing) not to discriminate on the grounds of sex, race or disability.

Testing during employment

There are six main circumstances when testing may be carried out during employment:

- on promotion or transfer
- routinely or occasionally
- on a random and unannounced basis
- following an accident at work
- following an incident at work when the employee's behaviour leads the employer to have a reasonable suspicion or belief warranting testing (collectively known as 'due cause' or 'with cause' testing)
- as part of an employee's rehabilitation programme.

Employers may additionally wish to include a provision in their drugs and alcohol policy to allow them to search employees and their property on the same bases.

Testing existing employees is much more problematic than pre-employment screening. Employers have no right to conduct alcohol or drugs tests or searches without the consent of employees. Any attempt to do so without an individual's permission would constitute a criminal offence.

If employers wish to conduct drug and alcohol testing and they have not already made an express provision in the employment contract, they should seek to negotiate the change with employee representatives or seek the agreement of individual employees. Alongside this, a change to the disciplinary rules will be necessary if employers want to make it clear that refusal to undergo a test (or search) will constitute a disciplinary offence that could result in dismissal. You should note that the reasons for refusal would need to be taken into account as these might be valid on medical or religious grounds.

In Chapter 5 we pointed out that in the absence of a contractual right an employee who has been forced to submit to a test (or search) may decide to resign and claim constructive dismissal. Should such a case go to tribunal, then an employer would have to rely on showing that the request was 'reasonable in the circumstances', eg there were reasonable grounds for suspecting that the employee was in possession of drugs.

Handling the results

It is vital that screening and testing are carried out under conditions of strict medical confidentiality. This raises the question of who within the organisation should be informed if a test result is positive. Occupational health staff may be appropriate or you may decide to appoint a policy co-ordinator for this and other purposes, eg to handle complaints about the policy and its operation. In any event, managers should only be informed as to whether employees or potential employees are fit for work or not.

The screening/testing process itself also requires careful management and monitoring. A proper 'chain of custody' must be maintained. This ensures that:

- the samples tested are actually provided by the person being screened/tested
- the samples cannot be tampered with
- the laboratory analysis and interpretation is guaranteed to be accurate.

(Laboratories accredited by the National Accreditation of Measurement and Sampling Service are certified to meet these standards.)

Occupational health staff may be best placed to carry out the important role of feeding back the results of the test to the individual concerned. Prior to the test, clear information should have been given on what might happen in the event of a positive test result.

Finally, it is vital to natural justice that employees and potential employees have a right to appeal against positive test results. Testing laboratories will keep all positive samples for one year in case an independent analysis is required.

Shortcomings

Aside from the controversy surrounding the practice of screening and testing, not everyone is convinced of their effectiveness. False positive and false negative results may be produced and legal substances such as painkillers can distort the results.

One problem is that a test only measures the amount of alcohol or drugs in the body at the time of testing. This may be perfectly satisfactory for an incident-related breathalyser test where the employer simply wishes to know whether the employee concerned has been drinking. It would be less valid as a means of detecting alcohol-dependency. Another problem is that, as some drugs remain detectable for several days and weeks after consumption, the test results may not be able to throw light on whether an employee has used drugs at work or whilst off duty. The competency of the individual concerned to perform the job is the key issue, as demonstrated by the *Sutherland* v *Sonat Offshore* case.

Case-study	
Case-law	**Key learning-points**
Sutherland v *Sonat Offshore (UK) Inc* 186/93 (EAT) A control room operator on an offshore drilling rig was dismissed, without a formal hearing, after he tested positive for cannabis.	The EAT held that the employer had been entitled to conclude that the scientific evidence indicated intentional use of cannabis on offshore premises – as opposed to passive smoking or off-duty behaviour. The contract clearly stated that bringing drugs or alcohol onto the rig or reporting to work in a drugged or inebriated condition were dismissable offences. A formal hearing would have served no useful purpose so the dismissal was fair. Employers should note that this case is an example of the harsher view taken by tribunals of employees involved in drugs-related incidents compared with those involving alcohol misuse. Another possible outcome would have been a finding of unfair dismissal because of the procedural deficiencies but with a significant reduction in the level of compensation awarded because of the employee's contributory conduct.

Employers who do introduce screening and testing may experience other shortcomings, for example:

- They risk alienating the majority of the workforce who do not misuse alcohol or drugs.
- There are several ways in which individuals can 'cheat' the testing system. (There are dozens of guides on the Internet.)
- Individuals who know they are to be screened/tested may simply avoid alcohol and drugs for long enough to escape detection.

The decision on whether to introduce screening and testing must therefore be weighed up carefully and, once taken, must be agreed with the workforce.

Ensure agreement

Despite the shortcomings listed above, drugs and alcohol testing is becoming more commonplace in the UK though its use is still limited to a small proportion of organisations. While testing will never be a substitute for drugs and alcohol policies, it can be used as part of a broader approach that encompasses raising employee awareness with a means of enforcing the standards of behaviour and performance expected at work.

For a successful introduction, employers must ensure that they secure the agreement of the workforce to the principle of screening/testing via the following means:

- consultation with safety representatives
- negotiation with union and/or other employee representatives (if applicable)

- seeking the written consent of each individual for each test undertaken for each specified purpose.

Drug and alcohol testing needs to be carried out sensitively if it is to be effective in helping to achieve the broad aim of ensuring a safe and healthy working environment.

How do we go about developing a policy?

- The right to manage
- Key issues
- Auditing the workplace
- Raising employee awareness
- Consulting the workforce
- Designing the policy
- Drafting the policy
- Communicating the policy
- Providing training
- Revising related procedures and organisational practices
- References

The right to manage

The introduction of a policy on drugs and alcohol raises a number of issues. One aspect that has already been touched upon is that of managerial prerogative, ie the manager's right to manage. Regulating an employee's behaviour may be viewed as an infringement of his or her personal freedom but employers have been doing this for many years within the work environment. Hence the existence of job descriptions, company rule books and a wide range of employment policies. Employers have an inherent right to manage and to decide on acceptable standards of conduct and behaviour.

It should be noted that, in the case of an alcohol policy, managers are not seeking to interfere in the normal social drinking that employees may engage in outside of work. The aim is to make employees aware of sensible drinking limits and the circumstances in which drinking can adversely affect work performance. With regard to drug misuse, managers have even more justification for insisting on compliance with 'the rules' as a failure to do so may constitute a criminal offence.

With this in mind, we can now consider the good practice steps necessary for designing a drug and alcohol policy.

Key issues

There are a number of key issues that need to be addressed when developing a drug and alcohol policy. They include:

- health, safety and welfare
- legislation
- workforce views
- support from top management
- education, training and communications
- support for individuals
- the scope of the policy
- confidentiality
- implementation and enforcement
- monitoring and review
- timing.

With these issues in mind, organisations should take a step-by-step approach to the process of introducing a new policy and allow a realistic time frame for this. Discussion, deliberation and consultation take time. An organisation will be in a better position to defend a complaint of unfair dismissal if it has a well thought out and properly implemented policy on drugs and alcohol.

The stages that must be followed can be put under these broad headings:

1. auditing the workplace
2. raising employee awareness
3. consulting the workforce
4. designing the policy
5. drafting the policy
6. communicating the policy
7. providing training
8. revising related procedures and organisational practices
9. supporting individuals
10. implementing and enforcing the policy
11. monitoring and reviewing the policy.

Stages one to eight will be examined in this chapter. In Chapter 8 we will be looking at likely inclusions in a policy. Stages nine to eleven will be covered in Chapter 9.

Please note that these stages should be tackled roughly in this order but are not intended to be discrete – for instance, the stage of raising employee awareness, ie educating the workforce, will be a continuous one.

A useful starting point is the establishment of a working party, drawn from all sectors of the workforce. The working party needs to agree its terms of reference and is likely to contain representatives from management, trade unions, personnel, occupational health, health and safety and employees drawn from different levels and functions in the organisation. The working party will have varying roles to play in all of the following stages.

Auditing the workplace

You may remember that in Chapter 4 we highlighted the possible causes of drug and alcohol misuse, many of which were linked to workplace pressures. An audit should therefore be carried out to determine the prevalence of drug and alcohol misuse in the workplace. The IPD Key Facts sheet *Stress at Work* states that this should cover[1]:

- people
- processes
- environment
- culture and other influences.

The same source recommends an independent attitude survey to explore employees' feelings about their jobs, their managers and the organisation and a risk assessment to identify high-pressure jobs.

> Other data to be used in the audit include records and information on absenteeism, accidents and work-related illnesses, labour turnover, working hours, shift systems, grievances and disciplinary cases, productivity, disputes, customer satisfaction surveys, employee assistance programmes, early retirements, involvement, suggestion/idea programmes, performance appraisal systems, training arrangements and personnel policies.

These statistics will be more meaningful where links can be established with drug and alcohol misuse. Further, other factors such as those jobs that are safety-critical ones, the availability of drink at work, and social pressures to drink would also need to be identified.

Raising employee awareness

Like most new initiatives in the workplace, an effective implementation is unlikely without the support and commitment of top management. It may therefore be necessary for the person championing the policy, possibly a personnel or occupational health specialist, to present a case to senior managers. They and indeed the rest of the workforce should be informed of the risks attached to ignoring drug- and alcohol-related problems in the workplace as well as of their own legal responsibilities. Top management also have an important role in setting a good example to the workforce.

Having read the preceding chapters, you will already appreciate that employers nowadays really have no choice in this matter – a policy is crucial to the defence of court or tribunal claims. However, raising the awareness of managers and the workforce in general should engender a sympathetic positive approach to colleagues with drug- and alcohol-related problems and an appreciation of the importance of prevention and recognition.

You can raise the awareness of all employees by conducting an educational campaign, possibly as part of a general health promotion or stress management programme. Techniques include:

- the distribution and display of appropriate literature (see Chapter 10 for a list of useful contacts)
- articles in company newsletters and other in-house publications
- inclusion of this topic in team briefings
- talks by outside experts
- announcing the commencement of the consultation process with a letter from the managing director or chief executive.

Consulting the workforce

The next stage is to begin the crucial process of consulting with employees and employee representatives in order to gauge the views of the workforce. There are a number of means available to achieve this end, eg meetings, use of employee suggestion schemes, questionnaires, ballots, surveys and discussions with trade union representatives and other employee representatives. It would be advisable not to rely on any one method but to use a combination of them.

As part of the ongoing educational programme, employees could be asked to complete check-lists or quizzes on drug and alcohol consumption, so that they can test their knowledge and evaluate their own practices. Surveys could be geared towards identifying whether employees are concerned about drugs and alcohol in the workplace and what their concerns are.

Further, as was stated in Chapter 6, if employers wish to include screening and testing in their new policy, it is essential that they secure the agreement of the workforce.

The initial consultation stage is of crucial importance in helping to clarify misunderstandings and to shape a policy which best serves all of your employees and ensures compliance with legislation. At its conclusion, the working party should be ready to begin the next stage of policy design.

Designing the policy

In designing your drug and alcohol policy, the broad areas for consideration are:

- alcohol – a total ban or restrictions based on consumption/timing of consumption/the nature of the job?
- drugs – a total ban is necessary to comply with legislation, but which 'offences' would be deemed more serious than others?
- screening/testing – under which circumstances will this be carried out and what would be the consequences of positive results?
- disciplinary action – when is this applicable?
- employee assistance – when is this applicable and under what conditions?

A range of example options under these headings and several others are set out in Chapter 8. Your choice will depend on the aims of your policy, the nature of your business, any specific legislative regulations and whether your employees are in safety-critical posts, deal with customers or the general public or entertain clients.

You must also consider the scope of your policy, eg:

- Are contractors and visitors to the workplace expected to comply with all the restrictions? Who is responsible for informing them? How will the policy be enforced?

- Do special arrangements apply for off-site events such as meetings, training courses etc?

There are many decisions to be made regarding the design of your policy. Highlighted above are the main areas for your consideration. The crux of the matter is that you need to choose a reasonable policy, ie one which takes into account the results of the consultation process and your responsibilities as an employer.

Drafting the policy

On the assumption that your working party has completed all the necessary research and has explored a range of possible options, you are now ready to draft the policy. You must ensure that sufficient time is built into the process to complete this stage as the consultation process is still ongoing at this point. Once the first draft has been agreed, it would be sensible to present it to the board of directors, management, union and other employee representatives. This would enable any unforeseen difficulties to be aired and, hopefully, resolved.

Employees should also be kept informed of the progress being made via notice-boards, memos, e-mail etc. The wording of the final agreed draft document should be clear and unambiguous, so that anyone reading it understands:

- that everyone is responsible for making the policy work
- what will happen if the policy is infringed
- who is responsible for administering and maintaining the policy
- the exact terms of the policy.

Communicating the policy

Here we consider the measures that are necessary to ensure that the content and purpose of the policy are communicated effectively to line managers and to all employees. Many policies are formulated after extensive research and consultation, but do not work because this stage is forgotten. If the policy is not widely known then employees will not be able to comply with its terms and will fail to report other employees who infringe the policy. Further, enforcing the policy will be impossible if it has not been implemented consistently throughout the organisation. Effective communication is therefore of paramount importance.

So how should you publicise your new policy? Many organisations have chosen to do this by undertaking some or all of the following:

- announcing the policy with a letter from the managing director or chief executive
- carrying out a series of briefings for line managers (including union officials and other key personnel, if appropriate)
- displaying the new policy on notice-boards and providing copies to all employees
- reinforcing the message via existing communication mechanisms, such as team briefings and internal newsletters
- putting up posters and leaflets on notice boards and making available booklets, videos and computer games
- changing relevant procedures and staff handbooks to reflect the new policy
- setting up a procedure for individual consultation

and support, ie instigating employee assistance programmes
- including references to the policy in job interviews, offer letters and at induction.

Providing training

An important stage in this process is that of training. Directors, senior managers, personnel specialists, line managers, occupational health staff, health and safety officers and union/employee representatives all require training. This should, according to the IPD, include[2]:

- information on the nature of drug and alcohol abuse
- the effects of drugs and alcohol
- case-studies
- law on drug abuse and alcoholism at work
- help available
- basic interviewing and counselling.

The aim should also be to tackle any prejudices or misconceptions that may exist, particularly in relation to drug usage and fears regarding links to the HIV virus. Further, we have already mentioned the difficulties inherent in seeking to distinguish between the signs of drug and alcohol misuse and symptoms of other medical conditions. Managers should therefore be trained to behave with caution on such occasions and, where there are doubts, seek a medical opinion.

Revising related procedures and organisational practices

As we have already stated, the implementation of a new workplace policy constitutes a change to the contract of employment and must therefore be handled carefully. You should consider whether:

- to specify in the disciplinary rules and procedure which breaches of the drugs and alcohol policy will be regarded as gross misconduct
- to include specific provisions in the contract of employment on, for example, entitlement to sick pay for drug or alcohol dependency, paid time off to attend employee assistance programmes and employer rights to test or search employees on a random or 'due cause' basis
- the existing grievance procedure needs to be amended to include a specific reference to complaints about the new policy and its operation or whether a separate procedure should be agreed to resolve disagreement, possibly involving the appointment of a policy co-ordinator (see Chapter 9 for more information)
- revisions are necessary to the staff handbook and health and safety policy
- the performance appraisal scheme needs to be revised (see Chapter 9 for more information)
- to inform contractors and other visitors in separate notes.

With regard to associated changes to organisational practices, employers should, for example, ensure that alcohol-free drinks are available at social functions, provide a good quality canteen service (to encourage employees to stay on site during meal breaks), provide transport where employees are entertaining clients and, where alcohol is banned from the site, ensure that it is also removed from the board room and executive dining room.

References

1 IPD Key Facts. *Stress at Work*. October 1998. p5.
2 IPM Factsheet 20. *Anti-Addiction Programmes*. August 1989. p2. (No longer available.)

What should a drug and alcohol policy contain?

- ✔ Types of policy
- ✔ Suggested headings
- ✔ Example policy
- ✔ Reference

Types of policy

We discussed in Chapter 3 the pros and cons of having combined or separate policies on drug and alcohol misuse. Employers must first decide which of these choices they prefer. This will depend on their particular organisational needs and circumstances and the policy aims and objectives. Further they may want the policy to be an extension of the organisation's health and safety policy, to be part of a wider health promotion strategy or to sit alongside other related initiatives such as stress management.

Policies may also apply at different levels, ie on a company-wide basis or at the division or plant level (or even across an industry). In any event, the policy should apply equally to all employees, regardless of status or seniority. This is particularly important when we remember that problem drinkers or drug users are just as likely to be found in the boardroom as on the shop floor.

In Chapter 5 we discussed the appropriateness of using the disciplinary procedure in some instances of drug and alcohol misuse. Organisations such as the IPD, HSE, ACAS, TUC, ISDD and Alcohol Concern all recommend that workplace policies should be supportive rather than punitive in their intentions, ie:

supportive – employers take a sensitive and non-judgemental approach by encouraging those with a problem to seek help voluntarily at an early stage. The organisation should seek to educate the workforce and should state its intentions with regard to confidentiality, job security and individual support to assist this process.

punitive – employers enforce the policy and ensure all employees are aware of the consequences of breaching the policy.

Workplaces differ widely so it is not possible to devise a model policy to suit all situations. Invariably, organisations will need to include both of the above approaches while also needing to draw a clear distinction between the situations to which they apply.

In some organisations, where safety considerations are paramount, there will be a need for strict and specific rules and procedures including provisions for screening and testing. For example, in the Sutherland case (see page 48), the contract clearly stated that bringing drugs or alcohol to work or reporting for work under the influence of alcohol or drugs were dismissable offences. Other organisations will follow less rigorous tactics. For instance, alcohol will not be allowed to be brought or consumed on the premises but lunchtime drinking off site is permissible if it is of a moderate nature.

As with smoking policies, it is possible that organisations may decide on a combination of measures with more stringent restrictions in some work areas than others. The difficulty is that this approach may be more difficult to implement and the opportunity is lost for, say, senior managers to set a good example to the rest of the staff.

The experience of organisations in the railway industry bears this out. There are wide variations in the jobs performed by employees with some, such as train drivers, holding extremely safety-critical posts. Thus the level of tolerance could be lessened for, say, office workers and managers. The appendix contains the policy for one such organisation, Railtrack. Here the decision was taken to apply the restrictions on drug and alcohol consumption to all employees (though unannounced testing only applies to safety-critical jobs).

Off-the-shelf solutions are not advisable, but there are some useful guidelines for organisations wishing to design and implement workplace policies. ACAS provides separate check-lists of items to be included in alcohol and drugs policies, and these have been combined (and slightly adapted) on page 66[1].

With this check-list in mind, we will now move on to suggest some headings to be included in your new policy. As before, we will, for simplicity's sake, work on the

Checklist of items to consider when drawing up policies on drug and alcohol misuse at work

- ✓ The purpose of the policy
- ✓ A statement that the policy applies to everyone in the organisation
- ✓ The rules on drugs and alcohol at work
- ✓ A statement that the organisation recognises that a drug or alcohol problem may be an illness to be treated in the same way as any other illness
- ✓ A statement that the rules will apply to any contractors visiting the organisation
- ✓ The potential dangers to the health and safety of drug users and drinkers and their colleagues if a problem is untreated
- ✓ The importance of early identification and treatment
- ✓ The help available – eg from managers, supervisors, the company doctor, the occupational health service or outside agency
- ✓ The disciplinary position – eg an organisation may agree to suspend disciplinary action where drug or alcohol misuse is a factor on condition that the employee follows a suitable course of action. Where gross misconduct is involved, a drugs or alcohol problem may be taken into account in determining disciplinary action.
- ✓ The provision of paid sick leave for agreed treatment
- ✓ The individual's right to return to the same job after effective treatment or, where this is not advisable, to suitable alternative employment wherever possible
- ✓ An assurance of confidentiality
- ✓ Whether an individual will be allowed a second period of treatment if he or she relapses
- ✓ The provision for education on drug and alcohol misuse
- ✓ A statement that the policy will be regularly reviewed
- ✓ Whether the policy has the support of top management
- ✓ Whether employee representatives have been consulted.

assumption that you have decided to design a combined policy rather than two separate policies.

Suggested headings

We provide below some example statements under each heading to assist you in drafting and agreeing your policy.

Purpose of the policy, eg:

- to help protect employees from the dangers of drug and alcohol misuse and to encourage those with a problem to seek help
- to ensure that employees' use of either drugs or alcohol does not impair the safe and efficient running of the organisation, or result in risks to the health and safety of themselves, other employees, customers and the general public
- to comply with all relevant legislation in this area.

Objectives, eg:

- to promote a culture in which drug and alcohol misuse is discouraged
- to encourage employees experiencing problems with drug or alcohol misuse to face up to the problem and seek help
- to ensure that the image and reputation of the company is maintained.

Scope of the policy, eg:

- The policy applies to all employees at all levels.
- The policy applies to all employees and contractors working on company premises.

- The policy applies to staff and students within the university.

Support for the policy, eg:

- This policy has the full support of the senior board.
- This policy has been designed following a full consultation with employees and trade unions.

General principles (such as prevention, recognition, conduct v capability and confidentiality), eg:

- The organisation treats drug and alcohol dependency as a medical problem that requires special treatment and help rather than as a disciplinary matter. It also recognises that early identification is more likely to lead to successful treatment.
- The company promises to maintain the strictest confidentiality when dealing with individuals, within the limits of what is practicable and within the law.
- Employees seeking help will be allowed time off for treatment and every effort will be made to assist them in returning to good health and full efficiency.

The rules – see the list on page 69.

Education and training, eg:

- Education and training is an essential and ongoing part of the company's approach to the problem of drug and alcohol misuse in the workplace.

Example rules	
Alcohol, eg	**Drugs**, eg
Employees are not allowed to drink during working hours and this includes meal breaks and 'on call' duties.	No employee must possess, consume, sell or give away illegal drugs whilst on duty.
No employee is banned from drinking but the amount of alcohol in their blood whilst at work must not exceed the legal driving limit.	Drug possession/dealing will be reported to the police, without exception.
No employee should report for duty within eight hours of drinking alcohol.	No employee shall report for work while under the influence of drugs.
There will be no consumption of alcohol on company premises, other than at the sports and social club and at special events, eg retirement parties and training events, authorised by the appropriate director.	Employees in safety-critical jobs who are found to be under the influence of illegal drugs will be liable to dismissal, regardless of the circumstances.
Employees who are representing the company, eg by entertaining clients, will be required to use discretion and limit social drinking whether this is during or outside of normal working hours.	Employees on prescribed medication which may affect their ability to perform their duties must notify the occupational health unit before reporting for duty.

- Information and publicity about drugs and alcohol in the workplace are conveyed through pamphlets, posters and notice-board information.
- Guidance will be provided to managers to enable the policy to be effectively communicated and implemented.

- Managers will be trained to recognise the early signs of drug and alcohol misuse and in effective interviewing and counselling skills.

NB: organisations may choose to include literature aimed at raising employee awareness in the overall policy or to have a separate employee awareness statement, often introduced prior to the overall policy.

Staff appointment procedures, eg:

- Applicants for vacancies will be informed of the organisation's drug and alcohol policy in the advertisement, on the application form and at interview.
- An explanation and a copy of the policy will be provided to new employees at induction.

Employee assistance, eg:

- Where an employee admits to a drug or alcohol problem, current disciplinary proceedings will be suspended and every effort will be made by the organisation to assist that employee in a successful rehabilitation.
- Where an employee has been diagnosed as having a drug or alcohol problem, time off with pay will be allowed for counselling or other treatment.
- If an employee has successfully completed a course of counselling or other treatment and later relapses, the line manager must decide whether to permit another period of treatment or to invoke the disciplinary procedure.
- During a period of treatment, the occupational health staff will keep the manager up to date

regarding the employee's progress, the likely 'return to work' date and whether alternative employment needs to be considered. After the return, the occupational health unit and the manager will jointly review the employee's progress.

Alternative employment, eg:

- If an employee's work responsibilities are seen to be an obstacle to their recovery, then redeployment will be considered.

Disciplinary action, eg:

- The following incidents are considered to be serious offences warranting dismissal:
  (a) possessing, using or selling illegal drugs in the workplace
  (b) being convicted of any criminal offence connected with drugs, regardless of whether the offence took place inside or outside the workplace.
- Where disciplinary action is appropriate but the employee concerned has a drug or alcohol problem, this may be taken into account as a mitigating factor.
- Where employees refuse to accept that they have a problem of drug or alcohol dependency or refuse treatment or the treatment fails, disciplinary action will be taken. Management reserve the right to terminate the employment contract.
- The employer is entitled to dismiss any employee who refuses to submit to testing or a search, without good cause.

- The employer is entitled to dismiss any employee who tested positive for illegal drugs or was found to be 'over the legal driving limit' with regard to the amount of alcohol in their blood in a random or 'due cause' test.

Screening/Testing, eg:

- The company runs a pre-employment drug and alcohol screening programme covering new employees at all levels.
- The company carries out random tests for drug and alcohol use on employees and tests employees who are involved in accidents at work or 'due cause' incidents. The employee will immediately be suspended with pay from work and will be required to sign a consent form before testing.
- The company is authorised to conduct drug and alcohol testing as follows on:
  (a) all employees – by providing written notification and at least 48 hours' notice or unannounced following accidents at work and 'due cause' incidents
  (b) employees in those posts that have been designated as safety-critical ones – by conducting unannounced (regular and random) tests, tests on transfer or promotion and tests following accidents and 'due cause' incidents.
- Employees must submit to searches of themselves or their belongings in the workplace during random security exercises and following 'due cause' incidents.

Responsibilities, eg:

- All employees are responsible for ensuring adherence to the policy and for reporting breaches of the policy.
- Managers should seek to identify the signs of drug or alcohol misuse and take appropriate but sympathetic action.
- Managers are responsible for ensuring that visitors and contractors are made aware of the terms of the policy.
- The personnel department is responsible for administering, monitoring and reviewing the operation of the policy.
- The policy co-ordinator acts as a point of contact for managers and staff seeking advice and, through the occupational health unit, provides access to specialist services.

Complaints/appeals, eg:

- Employees who have concerns about any aspect of the policy or its operation should initially submit their complaint, verbally or in writing, to their line manager and thereafter follow the formal grievance procedure.
- Employees will refer all their complaints about the operation of the smoking policy to the policy co-ordinator.
- If an employee or potential employee receives a positive test result, there is a right of appeal and the sample will be sent for independent analysis.

Review process, eg:

- Employees and trade unions will be consulted during the review of the policy and prior to the implementation of any amendments.
- This policy will be reviewed no later than x.x.xx and thereafter at 12-monthly intervals.

Timetable, eg:

- The consultation process has now been concluded and this policy will take immediate effect.

Example policy

An example alcohol and drugs policy is to be found in the appendix. You may wish to read it before moving on to the next chapter, where we consider how to avoid the pitfalls inherent in implementing any new policy.

Reference

1 ACAS Advisory Booklet. *Health and Employment.* Revised April 1997. pp34–35 and 42–43.

How do we make it work?

- ✔ Pitfalls
- ✔ Supporting individuals
- ✔ Implementing and enforcing the policy
- ✔ Monitoring and reviewing the policy

Pitfalls

There is a temptation in the introduction of any new policy within the workplace to think that the hard work has all been done by the final draft stage. Unfortunately, this is not so and just as much attention has to be paid to the later stages of:

- implementation and enforcement
- monitoring and review.

If this is not done, then there is a danger that nothing will change or, at best, implementation will be inconsistent. In the latter case, it would then be very difficult to police and enforce the policy as its terms and the consequences of breaching them would not be sufficiently well known to employees.

There are many potential pitfalls to be aware of when introducing a drug and alcohol policy and seeking to make it work. Some of the most common ones are:

- complacency in managers who fail to acknowledge the widespread nature of drug and alcohol misuse and the fact that it is likely to be a problem for some of their own staff
- a lack of consultation with employees such that the policy is perceived to have been imposed on them
- a failure to educate the workforce to the dangers of drug and alcohol misuse and to impress upon them the need to admit to their own problems or to report colleagues who they suspect are suffering from drug or alcohol dependency
- finding that the policy provisions are not as comprehensive as they should be and that many 'grey areas' exist where the rules are ambiguous
- conversely, finding that the policy is not flexible enough to cope with the wide variety of individual problems associated with drug and alcohol misuse.

In order to avoid these and many other pitfalls, you should follow the good practice guidelines set out in previous chapters and below. We will next examine the options available to employers for offering support and assistance to employees who suffer from drug or alcohol dependency.

Supporting individuals

Drug users and drinkers cannot be expected readily to adjust their patterns of behaviour just because their employer has decided to invoke a workplace policy. They will need advice, assistance and probably specialist support. Employers should also recognise that, when employees are undergoing treatment, relapses are common and they should have contingency plans in place for this eventuality.

There are various avenues through which the employer can offer help to employees. These are often collectively known as employee assistance programmes (EAPs) and they include the provision of leaflets on the dangers of drug and alcohol misuse, advice, counselling and help in fighting the dependency. Employers should note that specialist counselling skills are required for drug and alcohol problems. Where these are not available in-house, access to external experts should be provided. These may include hospitals, voluntary groups, counselling services and psychiatrists. (Some useful contacts are provided in the Chapter 10.)

There are many variations in organisational EAPs but most follow the same general principles:

- Employees are encouraged to seek help voluntarily from an appropriate source, eg the occupational health unit, staff welfare officers, the personnel department, company doctor or outside medical agency.
- Advice, counselling or medical treatment is offered confidentially.
- Time off with sick pay is allowed for treatment and the employee is given the same protection

and employment rights as other employees with health problems.

- Employees are treated sympathetically if they relapse and are given another chance if they agree to further treatment.
- Employees are made aware of the possible consequences if they refuse help or drop out of a recovery programme, ie the disciplinary procedure will be invoked.

Generally EAPs require employees to enter into agreements that commit them to certain actions:

- to follow the treatment and rehabilitation regime
- to abstain from the abused substance, ie drugs or alcohol
- to meet agreed targets in relation to work
- to consent to random alcohol or drug testing
- to agree to occupational health staff receiving progress reports from the treatment providers.

Implementing and enforcing the policy

We stated earlier that full consultation with employees and employee representatives is good practice in terms of the legislative requirements and employee relations.

With regard to implementation, it is likely that the personnel, occupational health and health and safety departments (if they exist) will be involved in this process. It may be necessary to nominate a member of the senior management team to take responsibility for the effective operation of the policy. It is also advisable for one person, possibly the policy co-ordinator, to maintain a database of information on the facilities available locally for the diagnosis and treatment of drug and alcohol dependency.

There are two aspects involved in enforcing the policy:

- dealing with infringements
- dealing with complaints.

With regard to the first aspect, we saw in Chapter 5 that it is imperative for a fair procedure to be followed, whether it is a disciplinary matter or a capability issue. An adequate system of monitoring work performance is therefore crucial to the smooth operation of a drugs and alcohol policy. Without it, early recognition of a dependency problem is much less likely and the chances of rehabilitation are reduced. Thus it may be necessary to review the organisation's performance appraisal system and associated management guidelines to ensure that they take account of all the possible causes of poor performance.

Normally line managers will be responsible for choosing the correct procedure following advice from the personnel department.

With regard to the second aspect, employers should actively encourage employees to make use of the grievance procedure if they have complaints about any aspect of the drugs and alcohol policy and its operation. (A separate specifically designed drugs and alcohol grievance procedure may be considered more suitable and may involve the appointment of a policy co-ordinator. In any event, it should follow the spirit and intent of the company's main grievance procedure.)

Grievance procedures are a means by which employees can officially raise complaints and seek redress. There are no absolute rules regarding the content of grievance procedures but the aim is to resolve the issue, to the satisfaction of all parties, as speedily and fairly as possible. You should note that poor handling of grievances can lead to employee discontent and, in certain circumstances, industrial disputes.

Thus grievances should be dealt with appropriately and within the terms of the procedure. These may include appeals against positive test results which we mentioned in Chapter 6. Even if a solution acceptable to all parties does not result, the fact that due attention was paid to the grievance should aid employers who may later need to defend their actions in, say, unfair dismissal claims.

Monitoring and reviewing the policy

The personnel department is usually responsible for administering employment policies and procedures. Thus they should ensure that the policy has:

(a) been integrated into the information (written and oral) given to prospective and successful candidates in adverts, at interview and during the induction process
(b) led to revisions to associated procedures such as the disciplinary and grievance procedures, the performance appraisal scheme guidelines, the health and safety policy and the staff handbook, so that contradictory documentation does not exist to cause confusion.

It is also crucial to monitor the operation of the policy to ensure that the rules are clearly understood and complied with. Where infringements do occur, these must be dealt with (and be seen to be dealt with) consistently and fairly across all staff. Where the organisational circumstances change, the policy should be reviewed accordingly.

Statistical evidence that may be useful in monitoring the success of the policy includes:

- any decrease in sickness and absence levels (especially where a link can be established to drug and alcohol problems)
- any decrease in accident rates
- the number of drug- and alcohol-related disciplinary cases
- the number of employees who have voluntarily sought help
- the number of drug- and alcohol-related problems that have been reported by colleagues.

The new policy, as well as being constantly monitored, should also be periodically reviewed through consultation with the employees and/or their representatives. The aim is to judge the policy against its purpose and objectives. Where there is a shortfall, amendments may need to be made. Specialists in occupational health and health and safety may be best suited to managing this process. This could include conducting annual health and safety surveys and audits of the opinions and attitudes of employees and managers. Information can be gathered on:

- the clarity of the rules contained within the policy
- the general awareness of the dangers of drug and alcohol misuse and what sensible behaviour constitutes

- whether the climate is conducive to self-referral
- the ability of managers and employees to cope with their responsibilities under the policy.

In this way, difficulties and problems can be brought to the surface and, hopefully, resolved. This should minimise employee discontent and lead to a happier and healthier working environment.

What else do we need to know?

- ✔ Further reading
- ✔ Useful contacts
- ✔ Employee assistance

We hope that the information provided above has convinced you of the need to establish a drug and alcohol policy in your workplace and that the good practice guidelines will help you to develop and implement a policy suited to your organisation's needs. Before starting this process, you should note that there is a lot more help available to you. Listed below are some suggestions for further reading and useful contacts.

Further reading

ACAS Advisory Handbook. *Discipline at Work*. Revised February 1999.

ACAS Advisory Booklet. *Health and Employment*. Revised April 1997. pp29–43.

Health Education Authority. *Don't Mix It! – A guide for employers on alcohol at work*. 1998.

Health and Safety Executive. *Drug Misuse at Work – A guide for employers*. 1998.

IDS Study. *Alcohol and Drugs Policies*. No. 652. August 1998.

ISDD and Alcohol Concern. *Drink, Drugs and Work Don't Mix*. 1999.

IPD Key Facts. *Stress at Work*. October 1998.

IPD Reward Group Salary Survey. *Drugs and Alcohol in the Workplace*. 1998.

Useful contacts

Alcohol Concern. Tel. 020–7928 7377

The Advisory, Conciliation and Arbitration Service (ACAS) can advise on the employment relations implications of alcohol and drugs policies. There are 11 regional public enquiry points – consult your telephone directory.

'Drugs and the Workplace' campaign. Tel. 01206–391800

Health Education Authority Information Centre. Tel. 020–7222 5300

Health and Safety Executive – the Employment Medical Advisory Service advises on all aspects of occupational health. HSE Information Centre. Tel. 0541–545 500

Institute of Personnel and Development. Tel. 020–8971 9000

Institute for the Study of Drug Dependence (ISDD). Tel. 020–7928 1211

Standing Conference on Drug Abuse (SCODA). Tel. 020–7928 9500

Turning Point for advice on drugs. Tel. 020–7702 2300

Employee assistance

Alcoholics Anonymous. Tel. 020-7833 0022

Employee Assistance Professionals' Association. Tel. 0800-783 7616

Narcotics Anonymous. Tel. 020-7251 4007

National AIDS Helpline. Tel. 0800-567123

National Drugs Helpline. Tel. 0800-776600

What are the key points?

- ✔ The law
- ✔ Alcohol and drugs policies
- ✔ Conclusion

The law

- By virtue of the duties of care imposed on employers by statute and common law, they are obliged to take reasonable steps to ensure that their employees are not under the influence of drugs or alcohol in the workplace and that they do not pose a threat to the health and safety of others.
- There is little specific regulation of the consumption of drugs and alcohol in the workplace except for the Transport and Works Act 1992, which makes it a criminal offence for certain railway workers to be unfit for work through drugs or drink.
- Under the Misuse of Drugs Act 1971, it is a criminal offence for employers knowingly to allow

controlled drugs to be kept, supplied or produced on their premises.

Alcohol and drugs policies

- Organisations should have clear rules on the use of drugs and alcohol in relation to the workplace.
- Polices should apply to all employees (and contractors, where applicable).
- Consultation should take place with all employees prior to the implementation of a drug and alcohol policy. A realistic time frame should be allowed for this process.
- Organisations should seek to implement a policy which is tailored to the organisational setting and takes account of employer and employee responsibilities.
- Suspected cases of drug or alcohol misuse should be fully investigated, and a medical opinion sought, before deciding on a course of action.
- It is appropriate to treat drug or alcohol dependency as an illness and to deal with it under the capability procedure.
- In misconduct cases, it is essential to follow a fair disciplinary procedure.
- Policies should be designed to encourage employees with drug- or alcohol-related problems to seek help and to assure them that they will be treated fairly and confidentially.
- An employer can insist on job applicants undergoing drug or alcohol tests, but there is no

inherent right to test or search existing employees unless there is an express term authorising this in the contract of employment.

- In enforcing the policy, managers must be trained to recognise the signs of drug or alcohol dependency, to interview and to counsel employees.

Conclusion

The message for all employers should by now be clear. If you do not have an alcohol or drugs policy, then one should be initiated as a matter of urgency. If you do have one, then it should be reviewed as soon as possible in the light of the relevant legislation and case-law.

Appendix: example policy

RAILTRACK

Alcohol and Drugs Policy

1. Introduction

This statement sets out Railtrack's Policy in respect of any employee or contractor whose proper performance of their duties is, or may be, impaired as a result of drinking alcohol or taking drugs. It is supported by the Rule Book, Group Standard on Alcohol and Drugs, related Codes of Practice, Guidelines and readily available educational materials.

Railtrack has taken into account the Transport and Works Act 1992. Provided that employees adhere to the provisions of this Policy, they will normally be able to demonstrate compliance with the Act.

All persons concerned are to be made aware of this statement and become familiar with its contents.

2. The Policy

Railtrack will take all reasonable steps to ensure that employees or contractors are made aware of the contents of this statement, together with the relevant sections of the Transport and Works Act 1992 and the implications therein. Furthermore, as a responsible employer, Railtrack will have in place procedures to prevent, in so far as it is reasonably practicable, an offence under the Act and a monitoring process to measure the effectiveness of such procedures.

It is a requirement of Railtrack that no employee or contractor shall:

- Report, or endeavour to report, for duty, having just consumed alcohol or being under the influence of drugs
- Report for duty in an unfit state due to the use of alcohol or drugs
- Be in possession of drugs of abuse in the workplace
- Consume alcohol or drugs whilst on duty.

Railtrack will not tolerate any departure from these rules and will take the appropriate disciplinary action in the event of any infringement.

Railtrack also has a policy of assistance with the rehabilitation of staff who voluntarily seek help for alcohol or drug-related problems. Such staff must, however, seek assistance at the earliest possible opportunity. Subsequent discovery or a disclosure prompted by an impending screening will not be acceptable.

A programme of screening has been put in place. This includes procedures to:

- Detect the use of drugs by both existing and potential employees

- Detect the use of alcohol and/or drugs by any person(s) involved in a Safety-Critical or Key Safety Incident where there are grounds to suspect that the action of the person(s) led to the incident
- Detect the use of alcohol and/or drugs where abnormalities in behaviour (which may include a request for screening) prompts managerial intervention.

Railtrack will measure the effect and adequacy of this Policy and the monitoring process annually.

3 The Railtrack Policy on alcohol and drugs

The Policy is a positive response to a more demanding legal framework and to a growing national problem of alcohol and drug abuse.

3.1 Our duty

Railtrack regards its employees as its most important resource and has a duty to them. In turn, Railtrack and its employees have a duty to their customers and to the public in general.

The Policy makes demands on Railtrack and all its employees, on you and on your professional approach to your work.

3.2 Alcohol and drugs at work

A clear link exists between the abuse of alcohol and drugs and:

- Safety
- Reduced efficiency.

The majority
The majority of employees are highly committed and disciplined.

The minority
Some may think they can come to work while unfit through alcohol or drugs and escape detection.

It is in the interest of everyone that any breach of the Policy should be prevented or detected.

The legal framework
The Transport and Works Act 1992 influences the way Railtrack must deal with the abuse of alcohol and drugs.

The Act makes Railtrack responsible for taking appropriate measures to prevent the risk of incident.

4 What are the main points of the Policy?

Railtrack will take all appropriate steps to ensure that no employee reports for duty, or tries to report, while unfit because of alcohol or drugs, or consumes or uses them while on duty.

4.1 What does Railtrack's Policy aim to achieve?
Railtrack aims to:

- Prevent risks to employees, customers and the general public from abuse of alcohol and drugs by Railtrack and other employees
- Protect the health and welfare of Railtrack employees by offering rehabilitation and counselling to those with alcohol and drug-related problems
- Prevent the damaging effects of alcohol and drugs on good business performance

- Comply with the Transport and Works Act 1992

What does the Policy apply to?
Without exception, the Policy applies to:

- All Railtrack employees
- Al employees of other companies operating on Railtrack lines
- All contractors and others working on Railtrack premises.

This means that the Policy applies to:

- You
- Your colleagues
- Your supervisor and manager.

In fact, to everyone.

Railtrack will ensure that the Policy is applied equally and fairly.

4.2 Why does the Policy go further than the law?
Evidence shows that when people have even a small amount of alcohol in their blood, their ability, performance and judgement are impaired.

This creates unacceptable risks to Railtrack.

5 How can alcohol and drugs affect us at work?

Alcohol and drugs can affect our behaviour; how we perform everyday activities and our work.

Every one of our actions depends on messages from the brain. Alcohol and drugs can delay and disrupt these messages.

With drugs, this applies whether we take them by injection, inhalation or orally.

Alcohol and drugs have very serious consequences for safety and can damage you, your work and the company.

5.1 The long-term effects

The long-term effects of alcohol and how Railtrack can help employees with an alcohol or drugs problem are covered in paragraph 14 (*see page 105*).

5.2 The short-term effects of alcohol and drugs

Alcohol and drugs can interfere with our:

- Co-ordination and the ability of our brain to control eyes, hands and feet
- Reaction, speed and our ability to judge distance accurately
- Short-term memory
- Ability to make rational and well-considered decisions.

Mixing alcohol and drugs can produce unpredictable results and is extremely dangerous.

Be safe.
Know the limits and stay within them.

6 What kinds of drugs are we talking about?

6.1 Illegal drugs

Railtrack forbids all illegal drugs and substances, such as:

- Heroin
- Cannabis/marijuana
- Cocaine
- Ecstasy
- Amphetamines.

6.2 Abuse of legal substances

Railtrack forbids all substances that are legal in themselves but are subject to abuse, such as:

- Glue
- Solvents.

6.3 Medicines

In addition, many medicines obtained with, or without, a prescription can affect performance at work. Examples are:

- Tranquillisers
- Anti-depressants
- Sleeping pills
- Some anti-histamines for hay fever
- Some medicines for coughs, cold and indigestion.

You must tell your doctor or chemist your job before you take any medicine and remember to advise your supervisor/manager that you are taking medication.

Remember, traces of some drugs can be detected days, or even weeks after use.

A trace of an illegal drug found during screening will lead to dismissal.

Be safe.
Don't abuse drugs & solvents.

7 Screening for alcohol and drugs

The purpose of screening is to:

- Detect any person who takes alcohol or drugs before coming to work
- Comply with the Transport and Works Act 1992.

Railtrack must exercise due diligence and do all that is necessary to maintain safety and ensure that an offence is not committed.

7.1 Drug screening – protecting your interests

The Chain of Custody

To make absolutely sure that a urine sample is secure and confidential, a Chain of Custody procedure is followed.

Trained Collecting Officer:

(1) Identifies donor
(2) Collects sample
(3) Transfers sample to container
(4) Labels and seals container
(5) Dispatches container in sealed package to laboratory.

Trained laboratory staff:

(1) Analyse sample
(2) Complete report
(3) Return report to Occupational Health Service
(4) Secure positive samples for one year in case the donor requests re-analysis.

7.2 Protecting your interest

Challenging the result

You have the right to challenge a positive result. The testing laboratory will keep all positive samples for one year in case you request independent analysis.

8 Pre-employment drugs screening

Candidates selected for employment will be screened for drugs. Any candidate who refuses to be screened will not be employed. Any candidate who tests positive will not be employed.

A leaflet on *Pre-employment drugs screening* is available to all applicants.

9 Transfer or promotion to Safety-Critical or Key Safety Post

If you are selected for transfer or promotion to a Safety-Critical or Key Safety Post, you may be screened for drugs, even if you already hold a Safety-Critical or Key Safety Post.

Before you take up the new post, you will be reminded that you must:

- Attend an Occupational Health Service
- Pass a urine sample for testing.

If the result is **negative**, both you and the appointing manager will be told.

If the result is **positive**, you will be dismissed.

If you refuse to be tested:

- You will not be appointed
- Disciplinary action could follow
- You could be dismissed.

A leaflet on *Announced and unannounced drug screening for persons transferred or promoted to or holding Safety-Critical or Key Safety Posts* gives further details.

10 For-cause screening

This is screening for either alcohol or drugs to find the cause of an incident or of suspect behaviour.

10.1 Post-incident screening

This is screening where there are reasonable grounds to suspect that your actions or omissions contributed to the cause of a Safety-Critical Incident and you will be:

- Tested for alcohol and drugs by the police, if they attend, or
- By a recognised agency, if necessary
- Removed from Safety-Critical or Key Safety work while waiting for the results.

Post-incident screening will give employees involved the opportunity of proving that alcohol or drugs played no part in causing the incident.

If the result of the drugs test is positive, you will be dismissed. If you fail the alcohol test, you will normally be dismissed.

If you refuse to be tested – disciplinary action will follow, and you could be dismissed.

If you are tested by the police, you:

- Could be arrested
- Must be prepared to give your manager a printed copy of the result of any alcohol test.

10.2 Screening as a result of behaviour or appearance

A manager or supervisor with reasonable cause to suspect that you are unfit for work through alcohol or drugs while on duty or when reporting for duty must arrange for you to be:

- Relieved from duty immediately, until you have been tested
- Tested for alcohol and/or drugs by a recognised agency
- Removed from Safety-Critical or Key Safety work while waiting for the results.

11 Unannounced drug screening

This screening applies to Safety-Critical or Key Safety Posts.

Is your job Safety-Critical or Key Safety? If it is, it will be identified:

- In your Safety Responsibility Statement
- In the Local Safety Policy Statement
- On internal vacancy lists.

If you have any doubts, ask your manager.

If your job is not Safety-Critical or Key Safety, you will not be subject to an unannounced drugs test.

A leaflet on *Announced and unannounced drug screening for persons transferred or promoted to or holding Safety-Critical or Key Safety Posts gives further details.*

11.1 Employees in Safety-Critical or Key Safety Posts will be selected randomly

Initially, five per cent of employees in Safety-Critical or Key Safety Posts will be randomly selected by computer.

If you are selected, you will be:

- Given up to 48 hours' notice
- Required to report for a test
- Required to pass a sample of urine for the test.

11.2 Waiting for the results

You will be expected to return to duty while waiting for the result of the test. If the result is positive, you will be dismissed.

If you refuse to be tested without good cause:

- You will be suspended from duty
- Disciplinary action will follow

- You will normally be dismissed.

12 What you risk if you disobey the Policy

12.1 A summary of the penalties

You will be dismissed, if you:

- Fail an alcohol test with 80 milligrams (mg) or more per 100 millilitres (ml) of your blood, or the equivalent in your urine or breath
- Test positive for drugs.

You will normally be dismissed, if you:

- Fail an alcohol test with 30–79mg or more per 100ml of your blood, or the equivalent in your urine or breath
- Refuse to take an alcohol or blood test without good cause
- Report or try to report for duty when unfit through alcohol or drugs
- Consume alcohol or drugs while on duty
- Possess illegal drugs while on duty
- Decline or discontinue an approved course of treatment for an alcohol or drugs problem, without good cause
- Seek help for an alcohol or drugs problem after you have been called for a test, and you fail that test.

If you are not dismissed:

- You will be disciplined and your personal record will always show this
- You will receive a final warning in writing if an

alcohol blood test shows that you have 30–79mg per 100ml of alcohol in your blood, or the equivalent in your urine or breath.

You are liable to prosecution if you offend against the Transport and Works Act 1992.

13 Know your drinks

It is important that you know how much alcohol you are drinking and that the amount of alcohol in drinks is measured in units.

13.1 What is a unit of alcohol?

The following drinks contain about one unit of alcohol:

- A half-pint of beer, lager or cider
- A single measure of spirits
- A glass of wine, sherry or aperitif.

13.2 How quickly does alcohol pass into the blood?

This depends on many factors, including:

- Gender – blood/alcohol levels in women are affected more than those in men
- Age
- Weight
- Metabolism.

13.3 How quickly do our bodies eliminate alcohol?

No matter how fast we drink, our bodies eliminate alcohol at the rate of about one unit per hour.

That means if an average man drank two pints of beer (four units) in one hour and then stopped, he would have about three units in his body after one hour, two units after two hours, and so on.

It would take about four hours to get back to zero.

13.4 Cold showers don't work

Cold showers, strong coffee and other sobering-up remedies have no effect on the amount of alcohol in your blood.

The alcohol test measures the amount of alcohol in your blood. You are responsible for the amount of alcohol in your blood.

It's very hard to work out exactly how much alcohol would be in your blood from the amount you drink.

13.5 How does the number of units of alcohol we drink relate to the amount in our blood?

The test measures the amount of alcohol in our blood in milligrams (mg) of alcohol per 100 millilitres (ml) of blood.

The more you drink, the more will find its way into your blood.

13.6 Discipline

If you have 80 mg of alcohol or more per 100 ml in your blood, you will:

- Fail the test
- Face dismissal, and
- Face possible prosecution.

If you have 30–79 mg of alcohol or more per 100 ml in your blood, you will:

- Fail the test
- Face dismissal, or

- Receive a final warning and a permanent entry on your personal record.

13.7 One milligram makes a big difference

If a test finds 29 mg per 100 ml in your blood, that is only just acceptable.

But, if the test finds 30 mg per 100 ml, you will have failed.

The difference between passing and failing is just one milligram of alcohol in your blood/alcohol level.

That equals one-fifth of a teaspoonful.

That's all it takes.

14 The long-term effects of alcohol

You should be aware how long-term heavy drinking can damage your health.

Railtrack is a caring organisation and is concerned about the health and welfare of all its employees.

14.1 What's the safe limit?

The safe weekly limits are 14 units for a woman and 21 units for a man. That's an average of two units a day for a woman and three for a man. Alcohol can damage a woman's health more easily.

14.2 Counselling and rehabilitation

If you think you have, or are developing, an alcohol or drugs problem, Railtrack has rehabilitation counselling and treatment programmes to help you.

You will be treated sympathetically and in confidence.

14.3 Get help now

Contact your local Occupational Health Service centre.

It will be too late to help you if you:

- Are involved in an incident, or
- Are called for a test, or
- Fail a test.

Don't leave it too late.

15 Code of conduct

To comply with the Railtrack Policy on Alcohol and Drugs, and to maintain the high standards of behaviour required, employees should avoid:

- Drinking alcohol during the eight hours before going on duty
- Drinking during meal breaks
- Drinking during paid on-call duty
- Wearing Railtrack uniform in licensed premises
- Using illegal drugs
- Having the smell of alcohol on their breath while on duty
- Being involved in an unexpected call out if they have any doubt about their fitness for work.